Royal Chapin Taft

Some Notes Upon the Introduction of the Woolen

Manufacture

Into the United States

Royal Chapin Taft

Some Notes Upon the Introduction of the Woolen Manufacture
Into the United States

ISBN/EAN: 9783337186180

Printed in Europe, USA, Canada, Australia, Japan

Cover: Foto ©berggeist007 / pixelio.de

More available books at **www.hansebooks.com**

SOME NOTES UPON

THE

INTRODUCTION

OF THE

WOOLEN MANUFACTURE

INTO THE

UNITED STATES.

BY

ROYAL C. TAFT.

PROVIDENCE:
SIDNEY S. RIDER.
1882.

This paper was read before the Rhode Island Historical Society,
April 18, 1882.

TWENTY-FIVE LARGE PAPER COPIES PRINTED.

PROVIDENCE PRESS COMPANY, PRINTERS.

PREFACE.

In January, 1871, upon the request of the "Rhode Island Society for the Encouragement of Domestic Industry," I prepared a paper upon the "Introduction of the Woolen Manufacture into the United States," which was published in the transactions of the Society for 1870.

The origin of this request was the following letter from Hon. Horace Capron, then Secretary of Agriculture, Washington, D. C., to a member of the Society, viz.:

"DEPARTMENT OF AGRICULTURE,
"WASHINGTON, D. C., January 18, 1870. }

"HON. JAMES DEW. PERRY, BRISTOL, R. I.

"MY DEAR SIR:— .... It has occurred to me to ask you to refer a matter of considerable importance to me, to your Society for investigation. The Hon. J. G. Dudley, in a paper read before the Historical Society of New York, claims that the first woolen factory built in the United States was by my father, Dr. Seth Capron, in 1809, in Oriskany, Oneida County, N. Y. Mr. Seth Newton Dexter, of that state, on referring to a letter written by Samuel Lawrence, in which he claims that honor for Rhode Island,—a woolen mill built in 1813,—says 'that Mr. Lawrence is widely in error. The first mill erected was by Dr. Seth Capron, at Oriskany, Oneida County, N. Y., with whom was associated DeWitt Clinton, Francis Bloodgood, Chancellor Platt, Smith Thompson, Stephen Van Renselear, Elisha Jenkins and others. Work was commenced in 1809, during embargo times, in anticipation of an act of incorporation, which was granted by the legislature in 1811.'

"If your Society should be pleased to take up this matter for investigation, I would be gratified to learn the result, because in so important a historical fact, credit should be awarded to whom it is due.

"I am Sir,

"Very respectfully,

"HORACE CAPRON."

The original paper, in order in point of time to meet the wishes of the Society, was necessarily somewhat hastily prepared and was slightly inaccurate in some minor details, and having always intended giving this matter further consideration, now in reviewing it, I am able to give additional and accurate information respecting the subject of the investigation.

I then reached the conclusion that Arthur and John Scholfield, were the first woolen manufacturers in this country. All subsequent enquiry has confirmed the opinion then expressed.

R. C. T.

PROVIDENCE, August, 1882.

WOOLEN MANUFACTURE.

The necessity for the introduction into our domestic economy of the industrial arts practiced and fostered in Great Britain, and so important to us as a nation, became more apparent, and found expression soon after the close of the war of the Revolution ; the obstacles placed in the way by her adverse legislation, but served to stimulate the enterprise of our fathers, who, having secured their independence, were now looking to the promotion of those industries in this country, which would best serve its interests, and render it independent of the old world. This desire was general throughout the country, but in no section was a progressive spirit more manifest than in New England. After the close of the war, many enterprises which had been undertaken came to an end, by reason of foreign competition.

George Bancroft, in "History of the formation

of the Constitution of the United States of America," says :

" The prospect of enormous gains tempted American merchants to import in one year more than their exports could pay for in three ; while factors of English houses, bringing over British goods on British account, jostled American merchants in their own streets."

" The people had looked for peace and prosperity to come hand in hand, and when hostilities ceased, they ran into debt for English goods, never doubting that their wonted industries would yield them the means of payment as of old. But excessive importations at low prices crushed domestic manufactures."

The heavy debt in which the colonies were involved, the lack of any sound financial system, and the absence of all laws regulating commerce, —allowing our markets to be filled to overflowing with the manufactures of Europe,—rendered the prospect for American manufactures in the future very discouraging.

Almost the first necessity of a people, is that for clothing, therefore the domestic manufacture of cotton and woolen fabrics was of the utmost importance, and its encouragement and protection received early consideration. As early as

February 3, 1781, Congress asked of the states as an "indispensable necessity," the power to lay a duty upon all imports, with no exemption except of wool cards and cotton cards and the wires for making them.

This first (though unsuccessful) scheme of duties on foreign commerce sought to foster American industry by the free admission of materials necessary to the manufacture.

July 2, 1785, the General Court of Massachusetts passed an act imposing a duty upon foreign manufactures, with the intent to encourage and protect their manufacture at home. This was the first protective tariff passed in the colonies. The legislature of Pensylvania passed a bill, September 20, 1785, to " protect the manufactures " of Pensylvania, laying a duty upon more than seventy articles, being only about two months behind Massachusetts.

The general court of Massachusetts appointed a joint committee, October 25, 1786, " to view any new invented machines that are making within this commonwealth, for the purpose of manufacturing sheep's and cotton wool, and report what measures are proper for the legislature to take to encourage the same." And upon the report of this committee an appropriation of two hundred pounds was made.

The importance of home manufactures is thus expressed in the Boston Gazette in 1788, viz. :

" Until we manufacture more it is absurd to celebrate the fourth of July as the birth-day of our independence. We are still a dependent people ; and what is worse, after the blood and treasure we have expended, we are actually taxed by Great Britain. Our imports help to fill her revenue and to pay the interest of a debt contracted in an attempt to enslave us."

In " A Topographical and Historical description of Boston, 1794," the writer says : " I would remark here, that many artists who arrive among us from abroad, are in poor circumstances, and are unable to set up manufactures for themselves. If such whose knowledge is competent to their profession, were assisted by wealthy citizens, they might become very beneficial members of society. By such means the various arts practiced in Europe, might in process of time be transplanted to America."

" As linens and woolens are very large articles of consumption, and carry out of the Commonwealth a large proportion of its specie, it would be well to pay attention to fabricating them here."

" Specimens that have been given of linen and

woolen cloths made here, demonstrate that we have manufacturers among us who are well skilled in making up the materials; and the number of them will increase by imigration from other countries."

"We are told that in 1667, a piece of woolen cloth was never dyed or dressed in England; it was improved by the skill of foreigners who came there; and that in a little more than a century the product was estimated at 16,800,000 pounds sterling, above 75,000,000 dollars per annum. Let us try what can be done in the United States."

The manufacture of cotton goods had become established, and was a recognized industry in 1790. The manufacture of woolens in the improved manner practiced in Europe, by machinery, had not yet been inaugurated; its introduction, without following its subsequent development, to any great extent, will be the object of this paper.

The imperfect manner of carding wool by the hand-card, when spun necessarily made uneven yarn, for which reason the cloth would be imperfect, the different parts shrinking unevenly in the process of finishing. This defect was remedied by the adoption of the machine carding, which mixed and carded the wool so perfectly that the different parts of the cloth would receive a uniform finish.

Therefore the use of the carding machine may be regarded as the initial point, the first step in the introduction of the woolen manufacture proper, or the fabrication by machinery.

The woolen manufacture had made so little progress at the close of the last century, that no contemporary recorded history exists, the little which can be found is largely traditionary, written long afterward from the recollections of those who have now passed away. It is fortunate in treating of this subject at this time that we have the living evidence of those, who if not actors, were witnesses to the introduction and of the operation of the first carding machine erected in this country.

In Felt's History of Ipswich, Mass., it is stated that the town of Ipswich granted land to John Manning in 1792, upon which to build a woolen factory, and subsequently made an additional grant, and in 1795, the town confirmed to Dr. Manning the land under the building. The building erected was 105 feet long, by 32 feet wide, being two stories high, built of wood.

The original design was to make woolen goods

and for a few years, broadcloths, blankets and flannels were manufactured; all the work of carding, spinning and weaving was performed by hand labor, but not proving profitable, cotton manufacture was substituted for woolen, and in 1800, operations entirely ceased.

In Coffin's History of Newbury, Mass., the author states that, "in June, 1794, the first incorporated woolen factory in Massachusetts was erected at the falls of the river Parker, in that portion of Newbury, known as Byfield parish. Most of the machinery was built in Newburyport, by Messrs. Standring, Armstrong and Guppy, Englishmen. In this year Benjamin Greenleaf and others were incorporated as the ' Proprietors of the Newburyport Woolen Manufactory.' .... The goods made there were broadcloths and flannels. While the factory was being erected a portion of the machinery was operated by hand."

In " Reminiscences of a Nonagenarian," by a lady of Newburyport, she writes: " The year I was seven years old the first incorporated woolen mill in Massachusetts was established at the falls on the river Parker, in the Parish of Byfield, in Newbury. The machinery for this factory was made in Newburyport, by Messrs. Standring,

2

Armstrong and Guppy, agents; the Messrs.
Scholfield and most of the operatives were English. The erection of this mill created a great
sensation throughout the whole region. People
visited it from far and near. Ten cents was
charged as an admittance fee. That first winter
sleighing parties came from all the adjacent
towns, and as distant as Hampstead and Derry
in New Hampshire. Row after row of sleighs
passed over Crane-neck hill, enlivening the bright
cold days by the joyous tones of their merry bells.
Never shall I forget the awe with which I entered what then appeared the vast and imposing
edifice. The large drums that carried the bands
on the lower floor, coupled with the novel noise
and hum, increased this awe, but when I reached
the second floor, where picking, carding, spinning
and weaving were in process, my amazement
became complete. The machinery, with the exception of the looms, was driven by water power,
the weaving was by hand. Most of the operatives
were males, a few young girls being employed in
splicing rolls. In a few years the first company
was dissolved, and the mill passed into other hands.

"The Scholfields were succeeded by Messrs.
Lees and Taylor. These gentlemen were also
English. New machinery, imported from Eng-

land for the manufacture of cotton goods, was put in. Mr. Taylor soon left, but Mr. Lees continued to operate the mill for several years.

"The establishment of this factory brought quite a revolution in the domestic manufacture of the neighborhood. For some time previous, in most families, hand carding had been discontinued, the wool having been sent to be converted into rolls, to the clothier's mills of Mr. Ben Pearson or Mr. Samuel Dummer.

"Lees and Taylor made arrangements by which this family carding could be done at their factory both cheaper and better than at the smaller mills.

"The introduction of cotton opened a new channel of industry. The weaving was still performed by hand; as the business increased this loom power was not sufficient to supply the demand for cloths. Their goods consisted of heavy tickings and a lighter cloth of blue and white striped, or checked, suitable for men's and boys' summer wear, aprons, etc. The tickings were woven by men on the looms of the factory, but much of the lighter stuffs were taken into families and woven on the common house loom."

Dr. Jedediah Morse, in the " American Gazetteer," printed at Boston in 1797, says : " A woolen manufactory has been established on an extensive scale in Byfield parish, and promises to succeed. "

In Bishop's " History of American Manufactures" is the following: " The first incorporated woolen company in Massachusetts erected a factory at the Falls of Parker river, in Byfield Parish, Newbury. The machinery was made in Newburyport. The stockholders were William Bartlett, principal, afterwards sole, owner, William Johnson, Nicholas Johnson, Michael Hodge, Joseph Stanwood, Mark Fitz, Mr. Currier, of Amesbury, Mr. Parsons, (late chief justice), Jonathan Greenleaf, James Prince, Abraham Wheelwright, Philip Coombs, and others. The English operatives by whom it was started, were Arthur, John, and James Scholfield, John Lee, Mr. Aspinwall, Abraham and John Taylor, John Shaw and James Hall, principally from Oldham and Saddleworth, England."

In reference to this enterprise at Byfield, is the following history, prepared from memoranda made by Nathan Scholfield, a grandson, now deceased, and from information furnished the writer by James and Thomas Scholfield, sons of John Scholfield, and by others of his descendants.

From which it appears that on the 24th of

March, 1793, Arthur Scholfield with John Scholfield and his family, sons of Arthur Scholfield,
who lived at Standich-foot, in Saddleworth, Yorkshire, England, sailed from Liverpool in the ship
" Perseverance " for the United States, where
they arrived the following May, at Boston. Upon
the landing of the two brothers, Arthur and John,
they introduced themselves to Mr. Jedediah
Morse,—author of " Morse's Geography and
Gazetteer,"—as being manufacturers, and well
skilled in the most approved method of manufacturing woolen goods in England. Arthur was
unmarried, John having a wife and six children,
was accommodated by Mr. Morse with a tene-
ment in Charlestown, near Bunker's hill, who
also provided for their immediate necessities,
and afterwards interested himself largely in their
behalf. Upon the return of the ship to Liverpool, the following letter of enquiry was directed
to the captain by Arthur Scholfield, their father,
viz. :—

" To Captain Delano Belonging the ship Caled
the Persivearance now liing at Liverpool.

" Honoured Sir we are no little Surprised and
Very uncesey that we have not yet Received a
letter from our Sons Arthur & John Schofield
who went on Bord Your Vessil to Boston in New

England, therefore Humble Desire you'l inform
them of it and let them know how unhappy we
are Concerning the'm likewise we sent a Box
after them by Your Brother who sailed in the
ship Caled the Dutiful Sons and we should be
glad to know whether the Received it or no And
if the are in want of anything from England it
shall be Sent them with all Speed and Humbley
desire You'l make it Conveniant Ether to write
to them or See them and inform them of all this
and I make not the least doubt But the will pleas
You for Your trouble and lethem know trades
of all sorts are Verry Bad and provishons of all
sorts very Dear things are strangely altered sins
the left England we are all well at present so I
remain Most worthy Sir with due Respects Your
Hum'e Serv't.

" Standige foot in Saddleworth,
" Yorkshire Aug 13th 1793
" ARTHUR SCHOFIELD."

After looking around for a few weeks, Arthur
and John determined to make a start in the manufacturing of woolen cloth, and on the 20th of
June took into copartnership with them a man
named John Shaw, a spinner and weaver who
had accompanied them from England, and at

once commenced the manufacture of woolen cloth by hand.

John Scholfield, being well skilled in the use of tools, built the first machinery himself, having completed a hand-loom and a spinning-jenny of forty spindles by the 4th of August of the same year. His books show that he paid £2,8,8, for lumber used in building the machinery, and that he charged the company for labor on the same £12,3,0.

He expended from his own funds for wool, £71,3,6, and on the 28th of October, sold from the first production of this loom, 24½ yards of black broadcloth, for £16,16, and 20 yards of mixed broadcloth for £12. All this work was done in the house occupied by them in Charlestown.

Mr. Morse was an interested observer of what was being done, and seeing that broad-cloth could be made to advantage in this country, and finding that Arthur and John understood the construction of machinery used in England, recommended them to some persons of wealth in Newburyport, who persuaded the brothers to remove to that place, for the purpose of starting a woolen factory with improved machinery, to be constructed under their supervision.

Upon their arrival at Newburyport, December 1, 1793,—taking with them the machinery built at Charlestown,—work was immediately commenced upon a carding machine, which was first put together in a room in Lord Timothy Dexter's stable, and there operated by hand for the purpose of showing its operation to parties desiring to engage in the enterprise. James Scholfield, living at Montville, Conn., now in his ninety-eighth year, was present at this exhibition, which he distinctly remembers.

This was in 1794, and was the first carding machine for wool made in the United States, and upon this machine were made the first spinning rolls carded by machinery.

Those interested in the enterprise, feeling thus assured of success, determined upon the immediate erection of a factory at Byfield ; the building was three stories high and one hundred feet in length, and was completed and started in 1795.

The first carding machine was made with a single cylinder, after which two double machines with two cylinders each were completed, and the three placed in the Byfield factory, where they were tended by James Scholfield, then eleven years old.

While the carding and other machinery was being constructed,—under the direction of Arthur and John,—the manufacture of woolen goods by hand was continued by them at Newburyport, as originally commenced at Charlestown, for their own account until October 12, 1794, when they sold their machinery to the company and removed to Byfield to superintend and start the new factory.

John Scholfield was employed as overseer of the weaving, and as agent of the company in the purchase of the wool. Arthur was overseer of the carding. John Shaw was employed as a weaver ; he worked in the factory for a number of years.

After remaining at Byfield about five years, John Scholfield, during one of his excursions into Connecticut and Rhode Island, purchasing wool, became acquainted with a valuable water privilege, at the mouth of the Oxoboxo river, in Montville, Conn., which he leased in 1798, from Andrew Tracy and wife, for the term of fourteen years. This lease runs to Arthur and John, is dated April 19, 1799, and includes the water-power, a dwelling house, shop and seventeen acres of land. In 1798-9, as soon as he could make arrangements to that effect, he, with his family and Arthur, left Byfield and removed to this place.

The business of the Byfield factory was carried on for a time by the company after Arthur and John Scholfield had sold their interest, but the company soon sold out to Lees and Taylor, who attemped to carry on the business, but shortly failed. It was subsequently operated for a time under the management of John Lees. John Lees and John Taylor were Englishmen, and were only operatives, employed as weavers in the factory while John Scholfield was agent.

In this enterprise where Arthur and John Scholfield were employed, the latter as agent of the company, to superintend the construction of the machinery and to conduct its business, we have the first instance of a woolen factory with improved machinery, erected in the United States, where the manufacture was successfully accomplished, all previous attempts having been unsuccessful, by reason of imperfect machinery.

Immediately upon the removal of Arthur and John Scholfield to Montville, they built a factory at Uncasville, a village in that town, which they put in operation as soon as completed.

Arthur, after continuing business here for a few years, being desirous of establishing himself elsewhere, sold out his interest in 1801 to John, and removed to Pittsfield, Mass.

John continued the operation of the Montville factory until 1806, when, owing to threatened difficulities with the owners of the adjoining land, regarding the right to the water-power used at the factory, and having purchased a factory property in Stonington, sold out to these parties, John R. and Nathan Comstock, leaving his sons James and Thomas to conduct the business until the termination of the lease.

This was the first woolen factory put in operation in Connecticut.

This property was owned by Nathan Comstock, Jr., as late as 1834, when he sold it to William G. Johnson, it being the present site of the " Johnson Dye Works," at Uncasville, the first privilege above the mouth of the Oxoboxo river.

It is most probable that the Scholfields imported a carding-machine from England while at Byfield, as, at the time of their removal to Connecticut, they took with them a carding-machine of which the frame, cylinders and lags were of mahogany. The sons of John Scholfield now living, were familiar with this machine, which they say was made in England, first put in operation

at Byfield, removed to Montville, and subsequently taken to the factory at Stonington.

Dr. Jedediah Morse always retained a lively
interest in the welfare of the Scholfield brothers,
and from the tenor of the following letter, written to John after they were located at Montville,
appears to have had financial transactions with
them.

" CHARLESTOWN, June 7, 1799.

" DEAR SIR.—" Your favor of the 6th ult'mo
was rec'd by Mrs. Morse while I was absent on
a journey to Phil'a. I take the earliest opp'y
since my return to answer it.

"The 18th of May, 1798, I paid Mr. How-
land 360 dols. I thank you however for your
intended indulgence. I had been wishing for a
long time to settle the balance with you, as I
knew there was a deficiency, Mr. Lyman having paid me a part through a friend after I had
drawn the order on him.

"Before I close this, I will endeavor to pro-
cure you and enclose the balance, it being
uneven money is not so convenient.

" My family are all in good health through a
kind Providence. Give our regards to your wife,

brother and children, and accept the same for yourself, from yours with esteem and affection.

" JED'H MORSE."

During the year 1806, John Scholfield bought a water privilege and Oil mill in Stonington, Conn., near Pawcatuck Bridge. This mill he filled with woolen machinery, and also built near by a factory building 30 by 40 feet, two stories high, which continued in his charge until 1812, when he returned to Montville, placing his son Joseph in charge, who operated the factory until 1834, when he sold the property to Orsmus M. Stillman. It is now standing and forms a portion of the Stillmanville Mills.

Joseph Scholfield was also connected in 1817, with others, in the ownership of a woolen factory in Dudley, Mass.

John Scholfield put into this Stonington factory, two double carding machines, twenty-four inches wide, two spinning-jennies, one with forty and one with fifty spindles, and a billy of thirty spindles; the jennies and billy were operated by hand, the carding machines by water power.

This was the second woolen factory in Connecticut, and the third place in which John

3

Scholfield had engaged in manufacturing,—first, at Byfield, in 1793–4; second, at Montville, in 1798–9; third at Stonington in 1806.

In 1813, John Scholfield purchased a factory and water privilege in Montville, located about four miles from the mouth of the Oxoboxo river, upon that stream; he enlarged the mill, putting in woolen machinery, and removed his family to this place, where he continued to reside during the remainder of his life.

This mill is now owned and occupied by his grandson, Benjamin Scholfield, who continues there the manufacture of the celebrated "Scholfield Satinet."

John Scholfield afterwards, in 1814, purchased a mill privilege at Waterford, on a stream which empties into Bolles Cove, three miles above New London, and there erected a factory, which was placed in charge of his son Thomas, who continued in that position until his father's death in 1820.

John Scholfield died February 28, 1820, aged sixty-two years, and was buried in Montville Cemetery. By his last will he gave free from all incumbrance, his three factory properties as follows: the one at Waterford, to his son Thomas; the one at Stonington, to his son

Joseph ; and the Montville property to his wife and younger children, which is now owned by one of his descendants.

In 1813 Thomas Scholfield, son of John, manufactured and sold the first piece of satinet made in Connecticut.   It was made upon a loom of his own construction ; the wool was from the farm of Christopher Greene, in Waterford, and cost one dollar a pound ; the warp also cost one dollar a pound.   This satinet was sold at three dollars per yard to Reuben Langdon, then a dry goods merchant in the building on the corner of Greene and State streets, New London.

The longevity of John Scholfield's family is something remarkable.   Of his six children, four are still living.   James, who resides at Montville, will be ninety-eight years old should he live to the coming September ; he enjoys good health and retains all his faculties in an unusual degree, and can read ordinary print without the aid of glasses.   Mrs. Mary Hinckley, a daughter, resides in Stonington ; she was ninety-five years old on the fourth day of February.   Thomas, residing at North Lynne, was ninety-one, and Isaac, residing at Noank, was eighty-one years old,—their birthday being the same,—on the twenty-first day of March.

The late Mr. Samuel Bachelder, in a communication to the Bulletin of the "National Association of Wool Manufacturers," vol. iv, refers to the former article by the writer upon this subject, and after quoting two different authorities in which Lees and Taylor are mentioned, in one of which it is stated, that Lees imported carding machinery from England, and put it in operation at Byfield in 1796, says : "Both of the above accounts agree in fixing the first operation of carding by machinery at Byfield, and in the establishment of a woolen factory there in 1794, and it is not improbable that Scholfield, Lees and Taylor are all entitled to some share in the credit of contributing this first step in the woolen manufacture."

From this detailed account from various sources of the origin of the Byfield factory, confirmed by the testimony of James Scholfield, who was then of an age to understand intelligently what was being done, it is shown that the Scholfields alone were the responsible managers of this enterprise, and that Lees and Taylor only appear after the Scholfields have severed their connection with the company, previous to this time they had simply been operatives in the factory.

There is much doubt as to Lees having imported woolen machinery from England for this mill.

In "Reminiscences of a Nonagenarian," she says: "The Scholfields were succeeded by Messrs. Lees and Taylor. These gentlemen were also English. New machinery imported from England for the manufacture of cotton cloth was put in."

Bishop, in his "History of American Manufactures," says: "Mr. John Lees, who had become proprietor of the woolen mill in Byfield, succeeded about this time (1805), in shipping clandestinely, from England in large casks labelled ' as hardware,' in charge of his brother-in-law, James Mallalow, a quantity of cotton machinery. . . . . . The machinery was erected in the factory building."

Mr. Bachelder's impression rests largely upon verbal information and tradition, which he admits is always liable to inaccuracy, particularly as to dates. John Lees subsequently re-appears as a cotton manufacturer at Holden, Mass., in 1822.

Arthur Scholfield remained with John at Montville about three years; he was married there on the sixteenth of April 1801, and in the autumn removed to Pittsfield, Mass., where he built a carding-machine, and commenced the business of carding rolls and manufacturing. He also built carding-machines and set them up for others to operate, as will appear hereafter. The remainder of his life was passed in Pittsfield, where he died March 27, 1827, aged seventy years and six months; he was buried in the rear of the Baptist church in that place.

The citizens of Berkshire county, having this industry brought to them thus early in its history, and the first of the textile industries to invite their attention, engaged actively in this branch of the manufacturing business, making it one of the principal centres for the woolen manufacture, and maintaining this supremacy down to the time of the war of the rebellion. It was said then that more woolen machinery was operated in this county than in any other within the state of Massachusetts. This statement would hardly hold good at the present time.

To the address of the Hon. Ensign H. Kellogg, delivered before the " Berkshire Association of Woolen Manufacturers," Feb. 22, 1855, is appended a sketch of the history of Pittsfield, as relating to the introduction of the woolen manufacture into that town, prepared by Mr. Thaddeus Clapp, 3d, wherein Arthur Scholfield appears as the prominent actor. From this sketch I have taken the following, viz.:

" Pittsfield was settled in 1752, incorporated in 1760, and in February, 1770, Valentine Rathbun started the first fulling mill. Mr. Rathbun purchased the property now owned by Messrs. J. V. Baker & Bros., and in a few weeks his fulling mill and hand-shears were in full operation. Stimulated by the extraordinary success of Mr. Rathbun (for he charged forty to fifty cents for fulling and finishing a single yard of cloth), Deacon Barber, in 1776, conceived the idea of establishing a rival concern in the north part of the town, and accordingly put in operation, near where the Pittsfield Company's mills are, an improved fulling mill, to administer to the wants of the neighborhood.

" From year to year, as the town increased, more clothiers' works were established, until the numbers of the profession became quite formidable."

In 1805, they had become so numerous that the idea of an association for mutual protection was suggested.

" A writer in the Pittsfield Sun, cf April 15, 1805, under the signature of 'Brother Clothier,' published an article, from which the following is an extract: ' If a society of clothiers should combine for the laudable purpose of investigating the natural quality of chemical liquids, and improve in making and dressing cloth, it would, in my opinion, be a society as useful and honorable to the country as a missionary or any other society whatever.'"

Mr. Clapp says: "Arthur Scholfield, the man who put in operation the first carding-machine, and manufactured the first piece of broadcloth in America, came to this country in 1789, with Mr. Samuel Slater, the father of cotton manufacture. Scholfield came to Pittsfield in 1800.

"The laws of England did not admit of the emigration of machinists, and therefore he took no tools with him, trusting solely to the power of his memory to enable him to construct the most complicated machinery. His memory was unusually tenacious, and being a good mathematician, he was enabled to enter into the nice calculations required in such a labor, but new and important obstacles came up, and he was obliged to return to England before he could perfect his carding machine. About the year 1801, his machine was completed, and we have his first advertisement in the Pittsfield Sun, of Nov. 2, 1801, as follows, viz.:

" 'Arthur Scholfield respectfully informs the inhabitants of Pittsfield, and the neighboring towns, that he has a carding-machine half a mile west of the meeting-house, where they may have their wool carded into rolls for twelve and a half cents per pound; mixed, for fifteen and a half cents per pound. If they find the grease, and pick and grease it, it will be ten cents per pound and twelve and a half cents mixed. They are requested to send their wool in sheets, as they will serve to bind up the rolls when done. Also a small assortment of woolens for sale.'

" The first broadcloth made in this country was by Scholfield, in 1804. This cloth was a gray mixed, and when finished was shown to the different merchants and offered for sale, but could find no purchasers in the village. A few weeks subsequently, Josiah Bissell, a leading merchant in town, made a voyage to New York for the purpose of buying goods, and brought home two pieces of Scholfield's cloth, which were purchased for the foreign article. Scholfield was sent for to test the quality, and soon exhibited to the merchant his private marks on the same cloth which he had before rejected. In 1808, Scholfield manufactured thirteen yards of black broadcloth, which was presented to James Madison, from which his inaugural suit was made. Fine merino sheep were introduced to this town about this time (they having been but recently introduced into the country, from the celebrated flocks of Rambouillet), and Scholfield was enabled to select enough to make this single piece, and President Madison was the first president who was inaugurated in American broadcloth.

"Some advertisements from the files of the Pittsfield Sun, of Scholfield's enterprise, will show what prices he obtained for his work, and how important his operations were regarded. I find on a day-book of his, broadcloth charged to individuals as early as 1805, and prices paid for weaving, from forty to sixty cents per yard.

"' PITTSFIELD FACTORY, April, 1805.

"' Good news for farmers, only eight cents per pound for picking, greasing and carding white wool, and twelve and a half cents for mixed. For sale, Double Carding-machines, upon a new and improved plan, good and cheap. Also, a few sets of cards made by the shakers [evidently hand-cards], and warranted good.

"' ARTHUR SCHOLFIELD.'

" ' PITTSFIELD FACTORY, 1806.

" ' Double carding machines, made and sold by A. Scholfield for $253 each, without the cards, or $400 including the cards. Picking machines at $30 each. Wool carded on the same terms as last year, viz.: eight cents per pound for white, and twelve and a half cents for mixed, no credit given.' "

Carding-machines having been purchased of Scholfield, were established at Lanesborough, in 1805 ; at Lenox ; at Curtis's Mills, in Stockbridge ; at the Falls, near the forge, in Lee ; at Mr. Baird's mills, in Bethlehem ; and in 1806, by Reuben Judd & Co., at Williamstown ; also by John Hart, in Cheshire, in 1807.

Two advertisements of the period, are as follows, viz.:

" FARMERS TAKE NOTICE.   CARDING-MACHINE. The inhabitants of this and the neighboring towns are informed that the subscribers have erected, about a mile west of Ezra Hall's tavern in Lanesborough, a carding-machine, at which wool of one color will be picked, oiled and carded for eight cents, and mixed for twelve and a half cents a pound.   The work will be superintended by a man who has served a regular ap-

prenticeship to the business; the strictest attention will be paid and every exertion used, to give
satisfaction to those who bring wool to their machine.

"BETHUEL BARKER, JUN., & Co.

"Lanesborough, May 10, 1805."

"CARDING-MACHINES. The subscribers, in addition to their old carding-machine, have lately
erected, at their mills near the Furnace in Lenox, a new and double machine, made and let by
Mr. A. Scholfield, Pittsfield, and by him warranted to be of the best kind; they now flatter
themselves they shall be able to give satisfaction
to all who bring their wool to their machines.
Strict attention will be paid by Mr. Perkin, who
with their present machines, can make as good
work as is made at any machine, or by any
workman in the country, (Mr. Scholfield having
relinquished the carding business). . . .

"WALKER & WORTHINGTON.

"Lenox, May 6, 1806."

"The first meeting to form a company for the
purpose of manufacturing fine cloth and stock-
ings, was held January 4, 1809, at Pittsfield.
The following is among the resolutions:—

" RESOLVED, That the introduction of spinning jennies, as is practiced in England, into private families. is strongly recommended, since one person can manage by hand, by the operation of a crank, twenty-four spindles."

This resolution respecting spinning-jennies, seems to imply but a recent knowledge of these machines, which is rather surprising, inasmuch as the spinning jenny was invented by James Hargreaves about 1767, and they were introduced from England into Philadelphia as early as 1775, and John Scholfield made one of forty spindles within three months from his landing; which would indicate a more general use of them in the country in 1809, than might be inferred from this resolution. A spinning-jenny of twenty-eight spindles for cotton was built in Providence by Daniel Anthony, in 1787.

The invention of the spinning-jenny by Hargreaves seems to have been the beginning of important improvements in woolen machinery in England; as from that time many inventions were successfully applied, which operated so powerfully upon the woolen business, that in the year 1800, it was found that the trade had increased three-fold in comparatively few years.

The statements of Mr. Clapp, relative to Arthur Scholfield, differ in several particulars from that of the sons of John Scholfield.

From their account he was associated previous to his removal to Pittsfield, in two enterprises with John Scholfield, one at Byfield in 1793-4; the other at Montville in 1798-9; in both of which broadcloth was made; consequently, the carding-machine erected in Pittsfield could not have been the first put in operation, neither was the broadcloth made there the first manufactured in the country. Also, it appears that he did not come from England with Samuel Slater in 1789, but with John Scholfield in 1793, neither did he return there during the time he was engaged in building his carding-machine; that was unnecessary, as the process of building carding machinery was already known to him, and obstacles could easily have been overcome by reference to the English machine he had left at Montville.

In consequence of the controversy between the United States and England in 1807, in reference to the "right of search," so called, an embargo was laid by Congress, on the 22d of December, upon all vessels within the United States. This measure was particularly obnoxious to the people of New England. They deemed it both impolitic and oppressive, and by

reason of this measure the large shipping interest of the United States was suspended until its
repeal.

Arthur Scholfield's business was seriously af-
fected by this measure, and he writes to his
brother John as follows, viz.:

"PITTSFIELD, July 11th, 1808.

"BROTHER JOHN Yours of the 4th June is
rec'd. You say you hardly know how you are
doing for there is an Imbargo laid last Dec'r,
and it still continues—the Imbargo is here too,
and likely to stay for what I see. It has swin-
dled me out of about 1500 dollars—for besides
what I shall loose by failures I have 22 Machines on hand besides Pickers—they were all
ingaged last summer, and if times had not turned,
should have had the money for them now. If I
had left Buiseness the spring before last it would
have been mnch to my interest but at that time
the Imbargo was not thought of, except by King
Jefferson and his party, and as they cant do rong
we must put up with it—I have often thought
you might have done Better by moving back into the Country, but as things are now there is
no doing anything anywhere—have not heard
from home a long time.

"ARTHUR SCHOLFIELD."

After the war of 1812, owing to the overwhelming influx of foreign goods, manufactures in this country became greatly depressed, and so continued for several years ; many who had started during the flush times of the war were obliged to suspend. Arthur Scholfield's business losses had been severe, and in 1818 he was advised by his friends in Pittsfield to make an application to Congress for relief in consideration for his services in the early introduction of the woolen manufacture to this country. He therefore wrote as follows to his brother John, at Montville, requesting his advice and counsel in the matter, as they were both equally interested, viz.:

"PITTSFIELD Apr. 20th 1818.

"BROTHER JOHN, Sir yours 20th Sept 1817 was duly recd, the reason I did not write sooner was I expected to have been able to pay Hicock without calling upon you again but finding it impossible I last week wrote to Isaac to know what situation that Legacy was in perhaps you have not heard that he had sold the goods to a man in Boston that had failed this he wrote me long ago and I thought by this time he might know something more about it but he writes me now that he has not rec'd a cent nor does he expect he ever shall but I don't wish you to distress yourself on my act tho Hicock is as needy and poor as any of us his family has been sick all Winter I was in hope our business would have been a little better by this time I have had a hard rub through the last winter but am in hope of doing a little better for the futer if we have our health—there is one thing I want to acquaint you with and have your opinion and advice about —i have been advised by my friends to apply to Congress by a petition as we were the first that introduced the woolen Business by Machinery in this country and should that plan be adopted I have but little hopes of success but they say if it does no good it wont doo any harm but at any rate I should like your opinion and advice about it the thing was suggested to me towards the last of the sessions, so that I had not time to write you on the subject and to do it on my own account I thought it would not be doing you justice as we were both equaly concerned, although I am personally acquainted with the member from this County and have faith to believe he would exert himself—Youl think of the thing and write me and accept of my best wishes for yourself and family.

" ARTHUR SCHOLFIELD."

John, in answer to this letter, wrote to his brother that he did not think the plan would succeed, and advised him to give it up. He also writes that his own business affairs are in a bad condition, owing to his having become bondsman for his son John, who had, by endorsing for other parties become involved in debt, and that he was in very feeble health.

To which Arthur replied as follows, viz.:

" BROTHER JOHN Yours of 17th Dec'r is rec'd —You have the same opinion about the legacy that I have (viz) that it is lost, but how it is I dont know—but as you say I have no idea of giving a receipt till I receive the money. I sent you an exact copy of what Isaac wrote to me, but as to the length of time you must lay that to me, for he wrote to me in the time of it and requested me to inform you whether I did or not I dont know.

" With regard to applying to Congress I have given that up for I am of your opinion that it wont succeed what gave me some hopes I was advis'd to it by a member of the Senet who is a very influential man in Congress but he is now out and I think tis best to drop it Your statement of your circumstances and what led to it is truly distressing but I hope you have not been so foolish as I was to become obligated for more than you are worth which was the case with me my situation as it respects property is worse than yours but thank God I enjoy my health as well as I ever did which I am sorry to hear is not the case with you.

"ARTHUR SCHOLFIELD."

Sylvester Judd, in his history of Hadley, says: "Carding-machines which were built in many towns after 1802, relieved women, who had before carded by hand. One was erected at North Amherst in 1803, one at the Lower Mills, in Hadley, in 1805, and one in North Hadley a few years later."

In "Lincoln's History of Worcester," mention is made of two enterprises: Joshua Hale, who began the carding of wool in the south part of the town in 1803, and Peter and Ebenezer Stowell, who commenced the weaving of çarpets and plaids in October, 1804, having six looms of their own invention and construction in operation; they also built shearing machines for wool.

Mr. Samuel Bachelder furnished for the history of New Ipswich, N. H., published in 1852, a sketch of the cotton manufacture in that town;

he there refers to James Saunderson, as connected in a collateral way with that industry,
and also, as an early woolen manufacturer in
that town.

James Saunderson came from a manufacturing district in Scotland to this country, in 1794,
and to New Ipswich in 1801, where he soon
afterward put in operation a carding-machine for
carding wool; this was the first carding-machine
introduced into the state. The woolen cloth of
household manufacture, which constituted the
principal clothing of the people, was imperfectly
made by reason of the primitive mode of carding. The advent of this machine created much
interest in this region, the inhabitants abandoned
the old process, and wool was brought from the
neighboring towns for a long distance to be
carded in this improved manner. He also carried on the manufacture of woolens, and from
1812 to 1814 his business was quite extensive.

Mr. Saunderson had also the skill (then almost
unknown in this country) of dyeing indigo blue,
by the same process as is now practiced in our
best manufacturing establishments.

The housewife could take her yarn to the
dye-house in the morning, have it dyed to a
beautiful and permanent color, ready to be car-

ried home at night; this was a matter of no inconsiderable wonder.

The skill of Mr. Saunderson afterwards proved to be of importance to the cotton manufacture of the town; in the production of colored fabrics he was employed to dye the yarn, and was subsequently employed by the Hamilton Manufacturing Company in skein-dyeing, soon after they commenced business at Lowell.

The first attempt at woolen manufacture in Rhode Island, was at Peace Dale, by Joseph Congdon and John Warren Knowles, who set up a carding machine in 1804, and soon afterward sold out to Rowland Hazard. This machine simply carded the wool into rolls which were put out to be spun by hand.

About 1812, Thomas R. Williams invented a power-loom for weaving saddle girths and other webbing, and probably in 1813, and certainly not later than 1814, these looms were started at Peace Dale. After they had been fully tested, Rowland Hazard purchased four of them for $300 each, and in 1814 or 1815, they were in successful operation.

The operation of power-looms at Peace Dale antedates those started in Judge Lyman's mill at North Providence, in 1817, by at least two years, and it is most probable that they were the first power-looms successfully operated in America, unless exception be made in favor of Francis C. Lowell, at Waltham, in 1814.

It is the opinion of James Scholfield, that the first application of water-power in this country for operating the spinning-jenny was by Mr. Hazard at Peace Dale. Isaac P. Hazard and Rowland G. Hazard, sons of Rowland Hazard, took charge of this business in 1819, and they with their successors in the family, have made many additions to the property, until, from this small beginning it has grown into the present extensive establishment of the Peace Dale Manufacturing Company, and has continued in the ownership of the family for nearly eighty years.

The late Hon. Zachariah Allen recently prepared for the writer an interesting sketch of the first attempt at woolen manufacture in this city. After speaking of the home manufacture by the hand-card and spinning-wheel, so universal before the introduction of machinery, Mr. Allen says :—

"The declaration of war with England in June, 1812, with a preceding embargo and non intercourse act, had advanced the price of manufactured cloths so excessively as to direct public attention to the branches of industry, of both cotton and woolen manufactures. An experienced manufacturer came to Providence, I believe from the west of England, and induced my brother-in-law, Mr. Sullivan Dorr, Samuel G. Arnold, Joseph S. Martin, Daniel Lyman, and E. K. Randolph, to form a company for the manufacture of broadcloths. This was the Providence Woolen Manufacturing Co. They commenced the erection of a large stone mill at the north end of Providence, with two wings and a dye-house.* A high pressure steam engine and cylindrical boilers were obtained from Oliver Evans, in Philadelphia, being the first steam engine for manufacturing purposes used in Rhode Island, as I believe. Apprehensions of the capture of it by British vessels induced the enterprising owners to arrange for the redemption of it by a liberal price, but it arrived safely.

"The cards were arranged on the lower floor of the centre building, the hand-looms in the wings and the spinning-jennies of forty spindles each on the upper floors. The shearing machines were of the Mussy pattern, used by hand, but were arranged by the ingenious manager, Mr. Sanford, to be operated by steam-power, with

---

* This mill now forms a part of the Allen's Print Works.

the cloth to traverse under the cutting-blades. A napping machine, made with pointed brass wires, arranged on a revolving cylinder, was newly invented, with adjustable parts to operate safely and efficiently. This machine and the fulling mills were placed in the basement. Mr. Sanford had a skillful dyer, Mr. Partridge. from the west of England, who was able to operate woad vats for blue dyeing. The colors he produced were highly admired, and the cloths were well made, and very durable; but the quality of the wool being somewhat coarse, most of the products were not of fine quality. During the war a quantity of Spanish wool was captured in prizes, which gave them a finer article at comparatively lower prices, and proved profitable for a time. They accumulated a large amount of broadcloths and refused an offer of eight dollars per yard, with the expectation of a further advance. But the arrival of the ship Bramble with news of an armistice signed by the Commissioners of the United States, at Ghent, put an end to all their hopes in the further manufacture of broadcloths. With the influx of foreign cloths of superior quality, the stock was closed out at a loss to the company of about $150,000, and the mill was closed.

" After the lapse of a few years the buildings were sold for a 'Print Works,' to Philip Allen, for which use they are occupied at the present day."

In 1822, Mr. Zachariah Allen erected a mill at Allendale, North Providence, for the manufacture of broadcloths. He there started the first power loom for weaving broadcloth operated in this state. Mr. Allen pursued the woolen business until 1839, using, as they appeared, the improved condenser for the carding machine, the improved English teazel cylinder, the extension roller (his invention, and first applied successfully at this mill), and other improvements in machinery. The first introduction of steam rolling, to give a gloss to the finished cloth, was at Allendale. In 1839, Mr. Allen sold the woolen machinery, and filled the mill with cotton machinery; it is still operated as a cotton mill by a member of his family.

Captain Abner Stearns, in 1805, purchased a water-privilege in West Cambridge, Mass., and erected a large building for the purpose of carding wool into rolls for hand-spinning in the families of the farmers. The whole of the second story was devoted to carding-machines and pickers. As there was no other carding factory in that section it was a great convenience to the farmers, who brought their wool for many miles around to be carded into rolls for spinning, and batting for hatters' use. He charged ten cents per pound for carding, and did a thriving business, often running the machines both day and night. In 1812, he erected another large building near his carding factory, where, with other machinery, he had a fulling mill and a spinning-jenny of seventy-two spindles; the yarn was taken elsewhere to be woven into broadcloth and then returned to the factory to be finished. During the war of 1812 he had a good business, but the peace of 1815, with the influx of British goods at low duties, rendered it so unprofitable that he sold out to James Schouler, a calico printer, of Lynn. These buildings were destroyed by fire, July 27th, 1875. Capt. Stearns was an ingenious mechanic, and an upright, enterprising citizen; upon selling out his manufacturing property, in 1816, he removed to his old homestead in Billerica, where he died in 1838.

In Hill's History of Mason Village, N. H., mention is made of an enterprise started by John Everett, and of his having erected a carding and fulling mill soon after 1800; there is no confirmation of this early date from any other source.

Mr. Bachelder places the date as 1810, when Mr. Everett commenced the manufacture of woolens; he erected a mill on the south branch of the Souhegan river. His first business was the fulling and dressing of woolen cloth that had been spun and woven in families; he afterwards manufactured satinet. In 1815, he paid twelve cents per yard for weaving.

According to Potter's History of Manchester, N. H., a project was started in 1809 and consummated in 1810, for the manufacture of cotton and wool at Amoskeag Falls, in Goffstown, N. H. The company was incorporated June 15, 1810, under the name of the "Amoskeag Cotton and Woolen Manufacturing Company." There is no evidence that the manufacture of woolens was entered upon; it was an unprofitable enterprise. The ownership changed several times; ultimately falling into strong hands, it formed the basis for one of the wealthiest manufacturing corporations in the country — the Amoskeag Manufacturing Company, Manchester, N. H.

The associations of the writer during his early life with the society and business of the town of Uxbridge, Mass., may excuse the somewhat lengthy history of woolen manufacture in that

town. Daniel Day, Joseph Day, and Jerry Wheelock, under the firm of Daniel Day and Company, built their first mill in 1810. Its size was
twenty by forty feet, two stories, containing a
carding-machine and picker, for the purpose of
carding rolls for home manufacture. In the
spring of 1811 they built an addition to the mill
of twenty-five by thirty feet, three stories high,
and in July put in a billy and jenny for spinning. In September they added a hand-loom;
early in 1812 they put in another loom, and during the year added three more, making five looms
in all. The picker (the mechanism used by the
weaver to throw the shuttle) was the same as in
use at the present time. It was operated by a
picker-string attached to the picker-stick held in
the hand, while the harnesses were operated by
the feet of the weaver.

The first weavers employed by Mr. Day were
English. Desiring to get more reliable persons,
he applied to Orsmus Taft (then a young man,
who was desirous of leaving the farm and of
learning the manufacturing business,) to go into
his mill to weave. He accepted the offer, at
what was considered by some of his friends and
the Englishmen, rather low wages. But he
thought, " let those laugh who win," and in

about a year he had charge of the weaving; now
Yankees generally took the place of the Eng-
lish. He always supposed that he was the first
American to weave satinet in Massachusetts.

Mr. Charles A. Wheelock, of Uxbridge, in an
appendix to the address of Hon. Henry Chapin,
delivered at Uxbridge in 1864, and published in
1881, gives an interesting history of the manufacturing business of that town, in which he refers to this enterprise of Messrs. Daniel Day &
Co., and says:

"Some three years since, in looking over some
old papers of my father's, which came into my
hands on the decease of my mother, I found a
receipt, of which the following is a copy:

"'UXBRIDGE, August 27th, 1811.
"'Rec'd of Jerry Wheelock seventy-five dollars in part payment for the picking and carding-
machine I have lately built and put in operation
in the shop of Mr. Daniel Day, in Uxbridge.
ARTEMUS DRYDEN, JR.'

"Here we have a glimpse of the beginning
of the woolen manufacture in Uxbridge, and, as
I believe, of the first woolen carding-machine
and picker built in Worcester county."

Mr. Dryden lived in the town of Worcester,

and from this early period to the present time, this has been one of the most important centres in the country for the manufacture of woolen machinery.

The introduction of cotton manufacture into this town was coeval with that of woolen, in 1810. The first movement was made at North Uxbridge, familiarly known as " Rogerson's," now the Uxbridge Cotton Mills. The billy and spinning-jenny were made by Arthur Scholfield, of Pittsfield, Mass.

The second attempt at woolen manufacture in Uxbridge was so characteristic of the time, when the country people of New England especially were ambitious for other occupations than farming, shoemaking and tavern keeping, and ready to coöperate in the then new and fascinating business of manufacturing, that Mr. Wheelock's history of this enterprise is given with considerable detail, viz.:

" The next attempt at woolen manufacturing was made by the Rivulet Manufacturing Company, which was incorporated in 1816, although the company was formed and buildings erected in 1814, and the business of manufacturing was begun in the winter of 1814 and '15. The capital paid in was $14,000 ; the shares were $500, each. It was agreed that no dividend should be paid until the expiration of eight years, a wise provision to make in this instance. As a matter of fact, no dividend was ever paid; and when the business was closed up, the stockholders received little, if any, more than half the amount paid in, and without interest.

" The original members of the company were, Daniel Carpenter, Samuel Read, Ephraim Spring, Alpheus Baylies, John Capron, Jerry Wheelock, Samuel Judson, Joseph H. Perry, Thomas Farnum, and Ezband Newell. The two last named persons, I think, soon surrendered their shares to the other members of the company. Daniel Carpenter was a merchant, and had been engaged in trade outside of an ordinary country merchant's trade, which well fitted him for the position he was now to assume—that of agent. Samuel Read was a farmer, hotel keeper, and owner of the privilege on which the mill was to be built. Ephraim Spring was also a farmer, and owner of real estate available for business purposes, besides having a son desirous of becoming a manufacturer in some of its branches. Alpheus Baylies was a farmer with sons who wished to become manufacturers. John Capron was a clothier by trade. cloth finisher and dyer,

whose proposition to the company will appear
by-and-by. Jerry Wheelock was a mechanic,
and one of the original Daniel Day Company,
and well acquainted with the construction and
operation of machinery, and with the management of stock, which would fit him for the place
of superintendent. Rev. Samuel Judson, the
Congregationalist minister, was, so far as I know,
the only man who might be considered a capi-
talist. He joined the company for the sake of
the profits from the investment, and a poor investment it proved. Joseph H. Perry was a
young man who came from Dudley, Mass., and
had money enough to take a share in the company and have an opportunity to learn a trade.
These men were all of moderate means, of sterling integrity, and good business qualifications
and intelligence.

"Surely such men were, and are now, the very
men and the only men fit to try the the coöper-
ative principle in business. This was a coöper-
ative association ; nothing more, nothing less.

" John Capron came to Uxbridge near the
close of the last century. The first mention of
his name that I have noticed on the town books,
is as one of the committee to superintend the
building of the school-houses in 1797. Ile had

acquired the trade of a custom clothier at the
Cargill mill, in Pomfret, Conn.   He purchased
the Col. Read estate and water-power, and set
up the business of finishing the cloth woven in
families in this vicinity.   This will account for
the following proposition :—

"'At an adjourned meeting of the Rivulet
Manufacturing Company, holden January 2d,
1815, at Capt. Samuel Read's, I made the following proposition to the meeting, in order to
join said company, viz.: that I would take
shares to the amount of $1,000, $1,500, $2,000,
or $2,500, and give my note to the company,
on interest; then to do the dyeing of all the
wool and the dressing of all the cloth for the
company, at the common price of doing the
same, till I had paid for as many shares as
they should choose I should take with them,
and that all charges for the same should be en-
dorsed on my note at the end of every ninety
days from the beginning till the whole be paid;
that I should then be entitled to the same value
of dyeing and dressing cloth for which said company are to pay me at the end of every ninety
days ; that is to say, that I shall do or cause to
be done, in manner as above stated, work to the
value of $5,000, in the whole.

" 'Then it was voted unanimously that I should take five shares, being the highest sum I had proposed, and in every respect agreeable to the foregoing proposition.

" ' JOHN CAPRON.

" 'Uxbridge, March 24th, 1815.'

" It is, therefore, easy to be seen why John Capron became a coöperator in this company.

" Artemas Dryden, Jr., made the carding-machine and picker for the company; and John and George Carpenter, of this town, built the billies and jennies; the first machinery built in this town, unless they had previously built a jenny for Daniel Day.

" The weaving was all done by hand-looms, and the goods were chiefly satinets, although some broadcloths and cassimeres were made.

" On the expiration of the contract with John Capron, the Rivulet Company put in finishing machinery, and, among other things, a shearing-machine with a revolving blade, or cutter, to be driven by power, then a recent invention by William Hovey, of Worcester."

As an instance of the prices paid for finishing woolen cloth, Mr. Wheelock gives a bill of Benjamin Cragin, of Douglas, against Daniel

Day and Company, of September 23d, 1813,
viz. :—

> " For Dressing 24 yds. wool cloth
>   N. Blue, at 25-100,   . . . $6.00
> For Fulling and Dressing 17½
>   yds. Satinet, at 20-100,   . .   3.40
>   ————
> $9.40"

About the entire cost of manufacture for
three-quarter yard wide goods, during the last
twenty years.

The other mills erected in this town previous
to 1830, will be briefly alluded to. That of
John Capron and Sons was built in 1820, or
1821 ; the Luke Taft mill, now Wheelock's, in
1825 ; and that of the Uxbridge Woolen Manufacturing Company, in the same year. This
was an incorporated company. The original
members of the company were Amariah Chapin,
Royal Chapin, Dr. George Willard, John and
Orsmus Taft. These men were all relatives,
and owners of the land on which the mill and
other necessary buildings, and tenements for the
employés, would stand, and of the most of the
land through which the canal leading to the
mill would pass. The Messrs. Chapin were
merchants and active business men, father and

son. The Messrs. Taft were brothers, both of them were manufacturers, and had been more or less engaged in the manufacture of woolen goods for several years.

In 1815, Samuel Slater (the father of cotton manufacture in the United States), in connection with Edward Howard, started a small woolen mill in the East Village, Webster, Mass., for manufacturing broadcloths and other woolens. This mill was destroyed by fire in 1820, when the purchase of a privilege was made at the South Village where the business was continued, forming a nucleus for the large establishment of the Slater Woolen Company.

It has been stated that this was the first attempt to manufacture American broadcloths. From what has preceded it is shown that this was not the first. A number of mills had made a previous attempt to manufacture broadcloth, and had succeeded; this is the only concern which has continuously made them, now for a period of more than sixty years.

Ezek Pitts commenced making woolen cloth in 1812, at the village of Millville, in the then town of Mendon, Mass.; his carding and spinning were done in an old building, awaiting the completion of his mill, which was in 1814, and

believed to be the first woolen mill on the Black-
stone river.

The growth of woolen manufacture was very
slow during the first decade of this century; the
limited supply of domestic wool doubtless had
its effect in repressing this industry. The growth
of wool was everywhere encouraged, but in 1810
the annual production had only reached 14,000,-
000 pounds.

From 1809 to 1815, woolen mills multiplied
rapidly throughout New England and the Mid-
dle States; at Oriskany, N. Y., in 1809; at Philadelphia, Pa., Pittsfield, Northampton, Watertown,
Uxbridge, and other places in Massachusetts,
from 1809 to 1812, and at Providence, R. I., in
the latter year. The State of New York granted
no less than twelve charters for woolen factories
during this year. The largest manufacturing
establishment for fine woolens at this time in
New England was the Middletown Woolen Manufacturing Company, at Middletown, Conn.

The manufactured product showed a corre-
sponding increase, in 1810 the annual production
had only reached the value of $4,000,000. But
the war of 1812 gave this industry the great advantage of our own market, freed from the com-
petiton of England, so that in 1815 the annual

production had reached a value of $19,000,000;*
and when we consider that previous to 1793, not
a woolen mill bordered the banks of our rivers,
not a yard of goods was made except those produced by the family from the hand-card and the
spinning-wheel, we can but look with surprise
upon the progress which this new branch of industry had made in this country to that time.

This investigation shows that John Manning
had land granted him by the town of Ipswich,
Mass., in 1792, upon which to build a woolen
factory, which grant was subsequently confirmed
to him in 1795, when the factory had been
erected.

The work done here was all performed by
hand, being no advance upon the method previously pursued.

This enterprise is presented as a representative
of several others, all earlier than the establishment at Byfield, and all using the more primitive
mode of manufacture before the introduction of
the carding-machine.

It also shows that Arthur and John Scholfield
came from England in March, 1793, with a knowledge of the process of manufacturing woolen
cloths, as pursued there; that they did during

---

* The Fleece and the Loom.   John L. Hayes, LL. D.

that and the following year erect and put into
operation wool carding-machines at Byfield,
Mass., which were the first erected; thus intro-
ducing the woolen manufacture into this country; that in 1798–9, they built a factory at Montville, Conn., and furnished that with the improved
machinery; also, that in 1801, Arthur Scholfield
left his brother John, and removed to Pittsfield,
Mass., where he erected the first carding-machine
introduced to that section of the country, and
followed the business of manufacturing woolen
goods with such success that in 1804 his broadcloths, consigned to the New York market were
sold in successful competition with the imported
article; while in 1808, he had made such sub-
stantial progress as to be able to make and furnish
the President of the United States with fine
American black broadcloth, for an inaugural suit,
this being the first (and perhaps the last) time
that a President of the United States has been
inaugurated in a suit made from cloth of home
manufacture; and also, that John Scholfield
engaged in his third enterprise in 1806, at Pawcatuck Bridge, in Stonington.

Of the six manufacturing enterprises with
which they were connected, four were earlier;
the first, fifteen or sixteen years previous to that

of Dr. Seth Capron, at Oriskany, N. Y., in 1809; referred to by Hon J. G. Dudley, in his paper read before the New York Historical Society, as being "the first woolen factory built in the United States."

Other factories were built soon after 1800; that of James Sanderson, at New Ipswich, N. H., and at Amherst, Hadley, Worcester, West Cambridge, and other towns in Massachusetts and in Connecticut.

The carding-machines erected at this period in New Hampshire, Massachusetts and Connecticut had their origin in the enterprise and skill exhibited by Arthur and John Scholfield, by their introduction and successful operation of this improved machinery for perfecting the process of manufacturing woolen goods; to them should be awarded the credit due to pioneers of this industry, and also due to them, as being the first successful woolen manufacturers in the United States.

www.ingramcontent.com/pod-product-compliance
Lightning Source LLC
Chambersburg PA
CBHW022008190326
41519CB00010B/1440